Monsters
Preschool Basics Workbook

This book belongs to:

_____

_____

Copyright © 2019 by KIDSFUN

All rights reserved. No part of this publication may be reproduced, distributed, or transmitted in any form or by any means, including photocopying, recording, or other electronic or mechanical methods, without the prior written permission of the publisher, except in the case of brief quotations embodied in critical reviews and certain other non-commercial uses permitted by copyright law.

Table of Contents

Number Tracing.

Match the Number.

Let's Count!

What a Difference.

More or Less?

Big vs. Small.

Compare sizes.

Shape.

The picture that comes next!

Number Fun.

0 Zero

Trace the number 0. Color the zero

0 0 0

0

Zero Zero Zero

Zero

0

Zero

0

Zero

Circle the square with _0_ image.

Color _0_ Christmas trees.

1 One

Trace the number and color picture.

1 1 1

1

One One One

One

1

One

1

One

Circle the square with _1_ image.

Color _1_ Christmas tree.

2 Two

Trace the number and color them.

2 2 2

2

Two Two Two

Two

2

Two

2

Two

Circle the square with _2_ image.

Color _2_ Christmas trees.

3 Three
Trace the number and color them.

3
Three
3
Three

Circle the square with _3_ image.

Color _3_ Christmas trees.

 Four

Trace the number and color them.

4 4 4

4

Four Four Four

Four

4

Four

4

Four

Circle the square with 4 image.

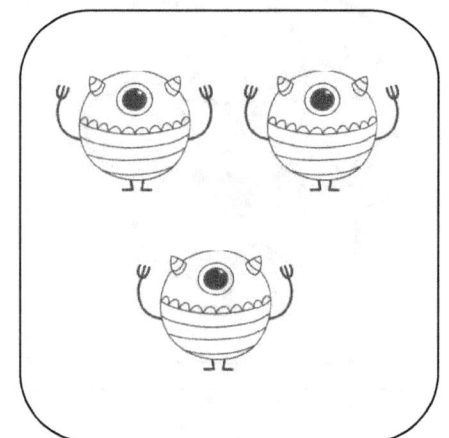 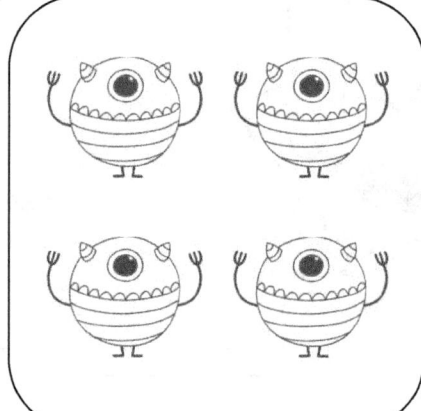 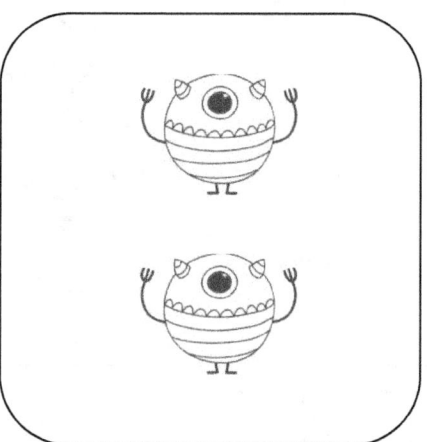

Color 4 Christmas trees.

5 Five

Trace the number and color them.

5 5 5

5

Five Five Five

Five

5

Five

5

Five

Circle the square with 5 image.

Color 5 Christmas trees.

6 Six

Trace the number and color them.

6 6 6

6

Six Six Six

Six

6
Six
6
Six

Circle the square with _6_ image.

Color _6_ Christmas trees.

7 Seven

Trace the number and color them.

7 7 7

7

Seven Seven Seven

Seven

7
Seven
7
Seven

Circle the square with 7 image.

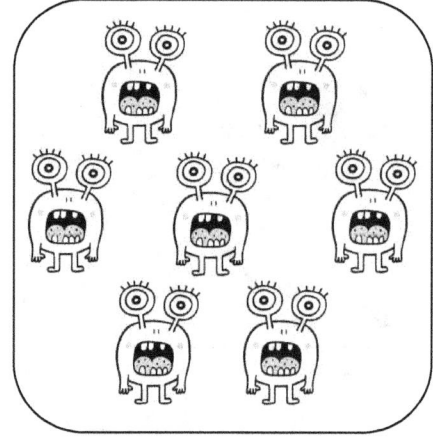 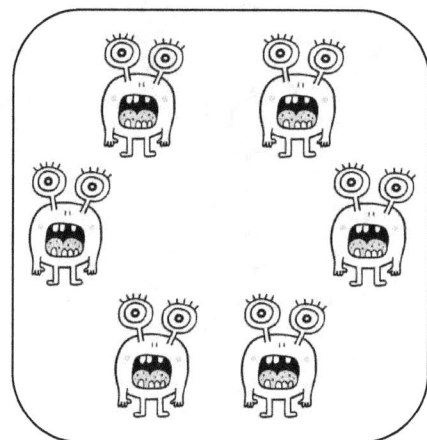 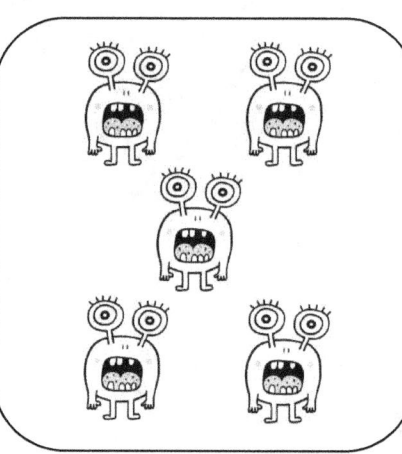

Color 7 Christmas trees.

8 Eight
Trace the number and color them.

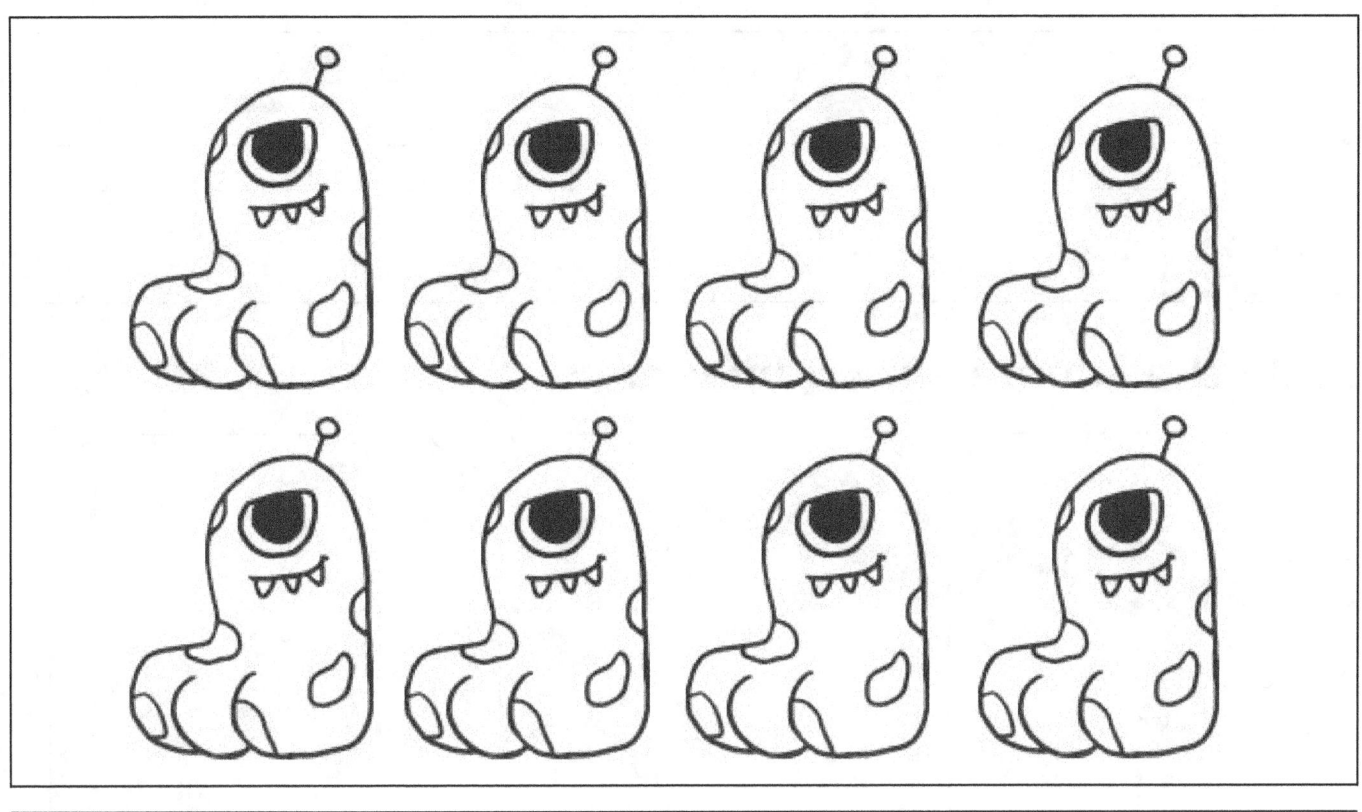

8

Eight

8

Eight

Circle the square with _8_ image.

Color _8_ Christmas trees.

9 Nine

Trace the number and color them.

9 9 9

9

Nine Nine Nine

Nine

9
Nine
9
Nine

Circle the square with _9_ image.

Color _9_ Christmas trees.

10 Ten

Trace the number and color them.

10 10 10

10

Ten Ten Ten

Ten

1 0

Ten

1 0

Ten

Circle the square with _10_ image.

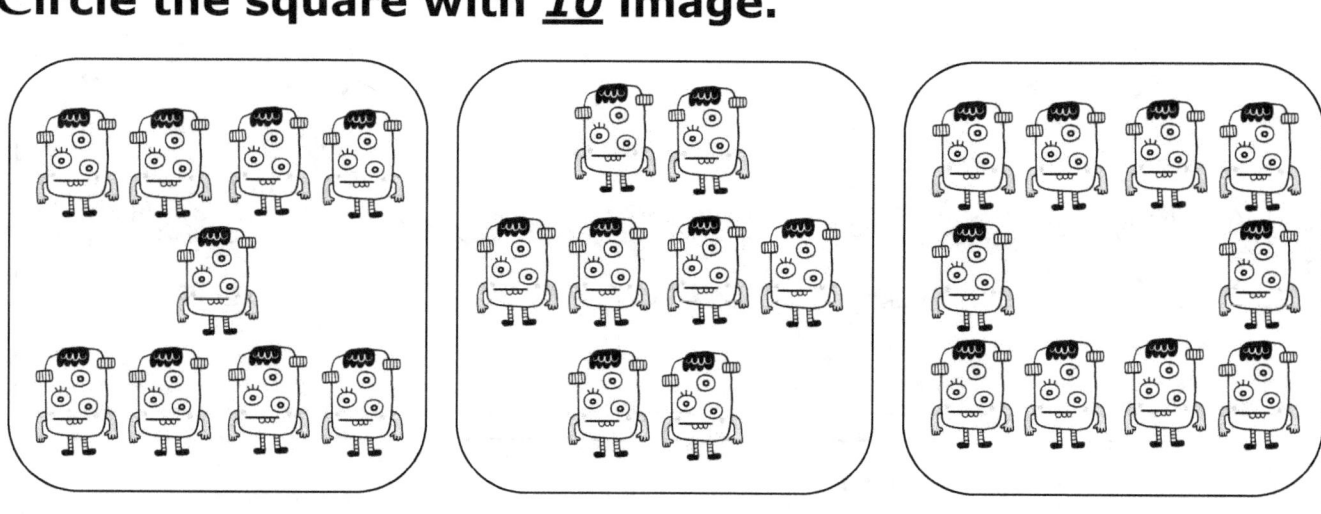

Color _10_ Christmas trees.

Color the monsters and Match the Number!

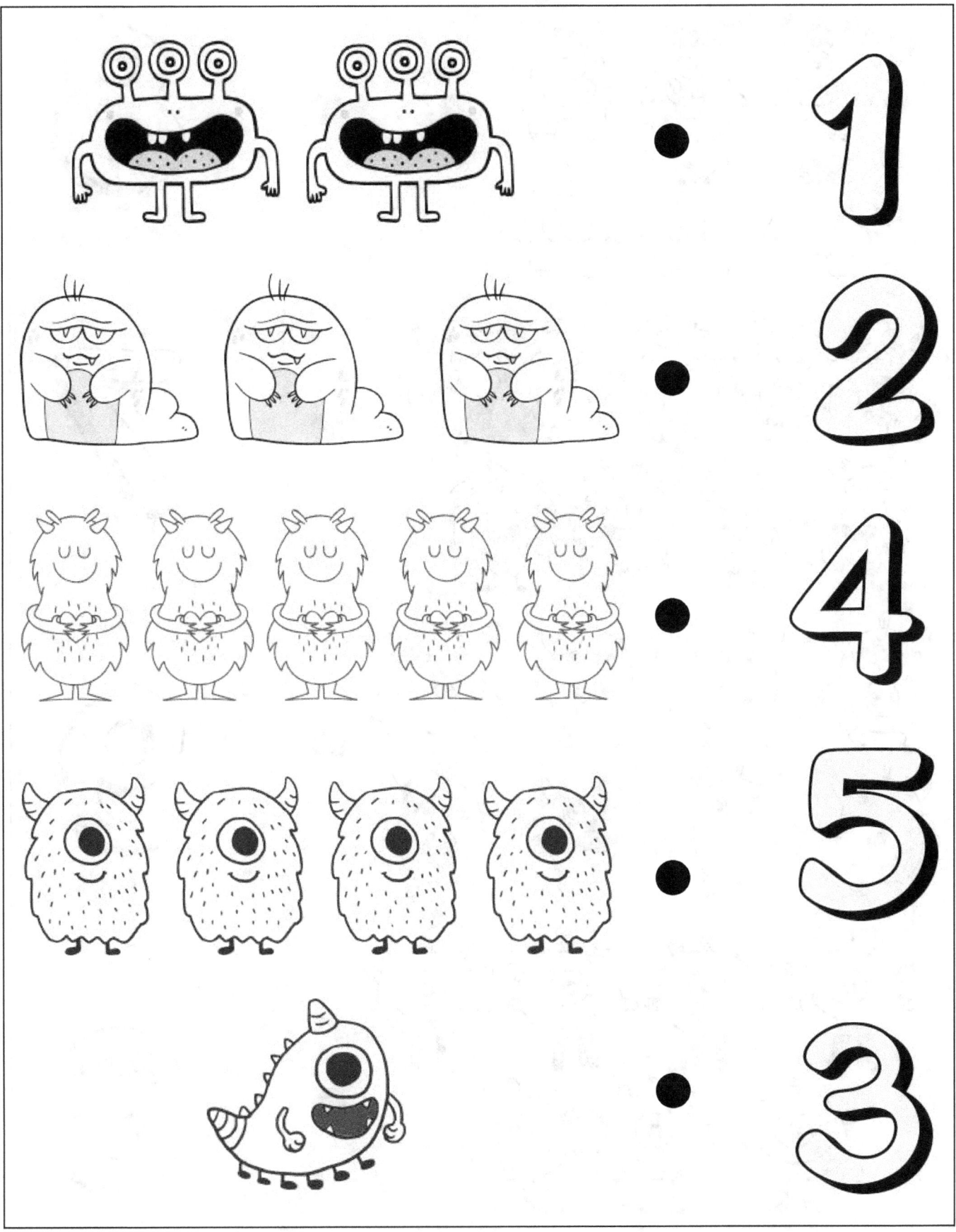

Color the monsters and Match the Number!

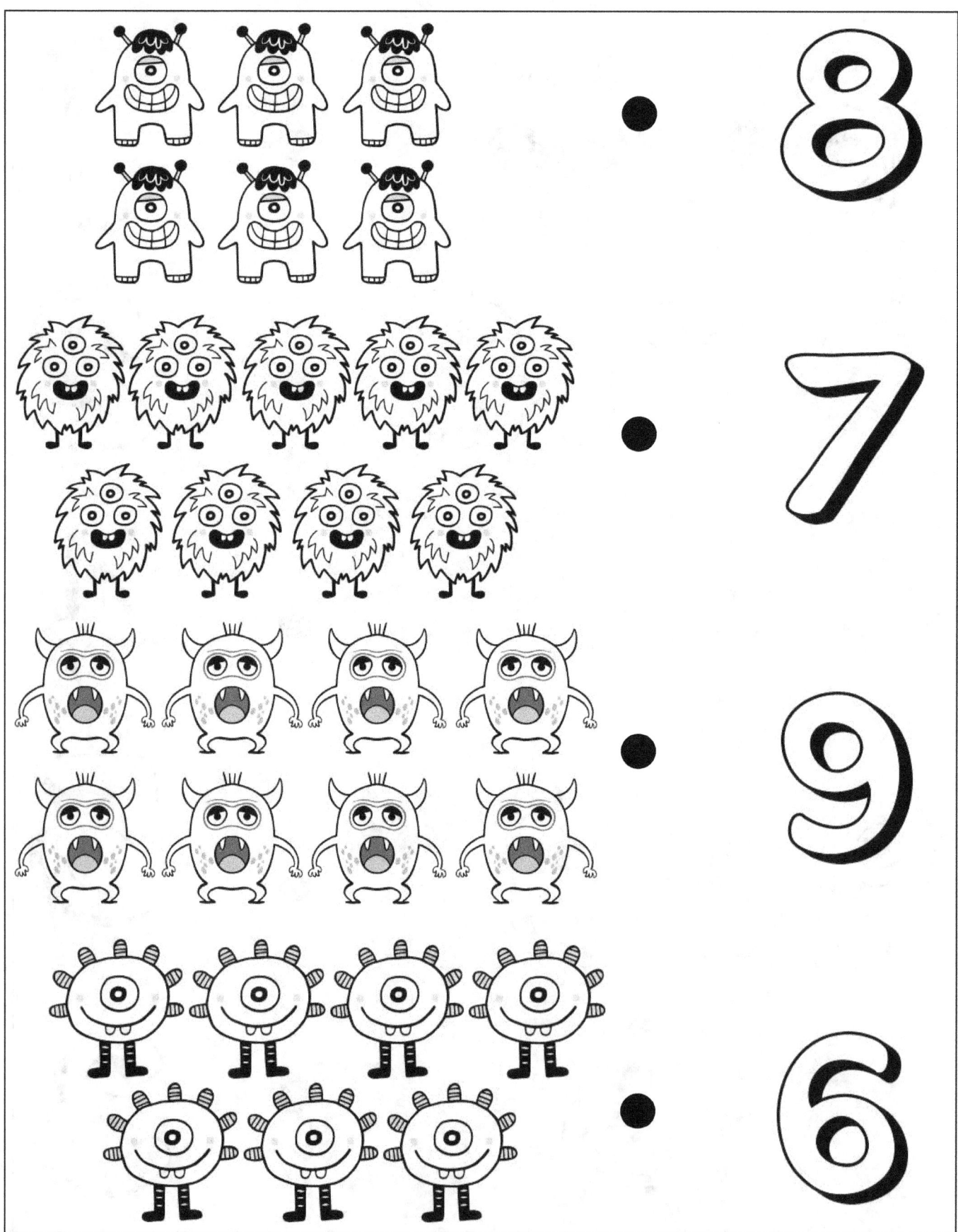

Let's Count!
Count each kind of monsters and fill your answers in below.

How many 🐮 ?_____

How many 🐐 ?_____

How many 👾 ?_____

Let's Count!
Count each kind of monsters and fill your answers in below.

How many  ? _____

How many  ? _____

How many  ? _____

Let's Count!
Count each kind of monsters and fill your answers in below.

How many 🧟 ?_____

How many 🧟 ?_____

How many 🧟 ?_____

Let's Count!
Count each kind of monsters and fill your answers in below.

How many  ?_____

How many  ?_____

What a Difference
Look at the pictures below and color the correct picture.

What a Difference
Look at the pictures below and color the correct picture.

What a Difference
Color in the pictures. Make them all different!

What a Difference
Color in the pictures. Make them all different!

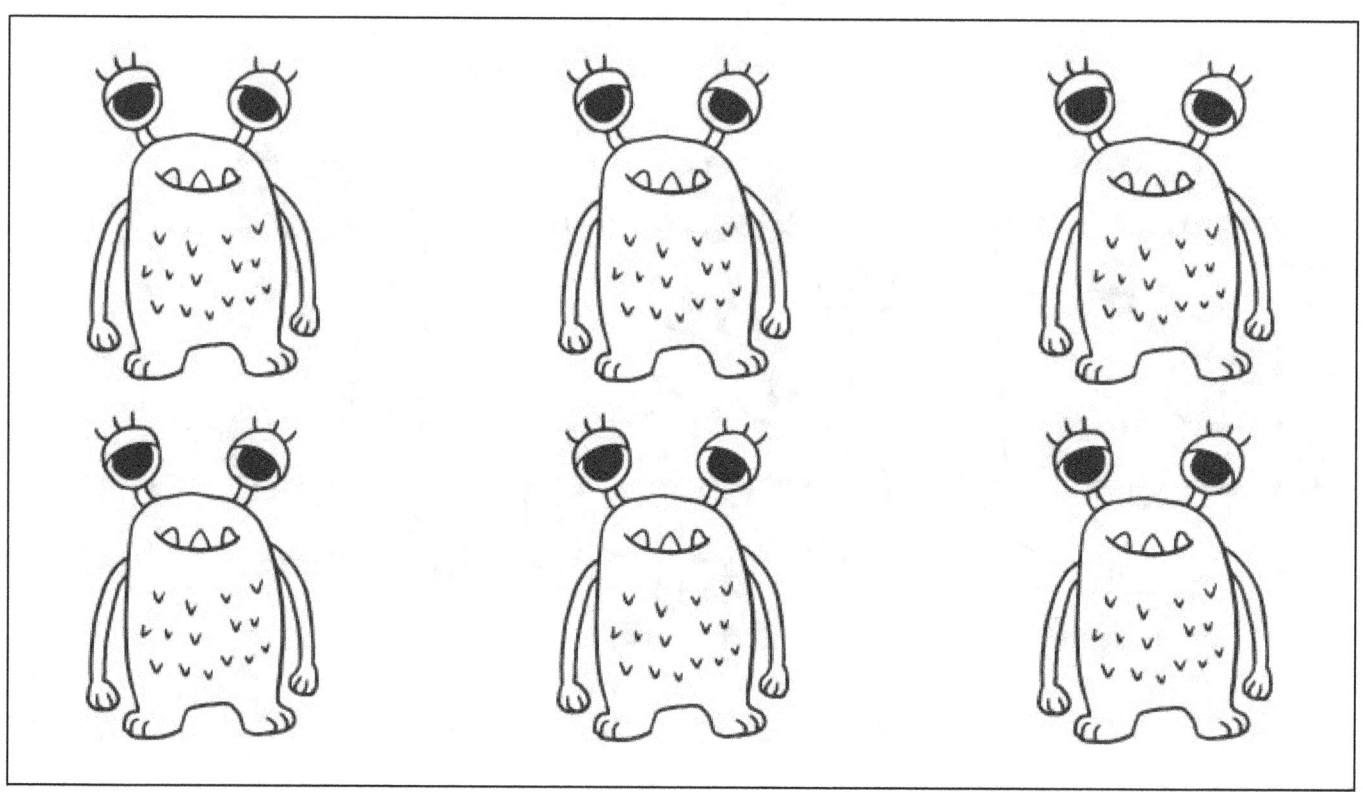

More or Less?

Look at the pictures in each box and circle the group that has **more.**

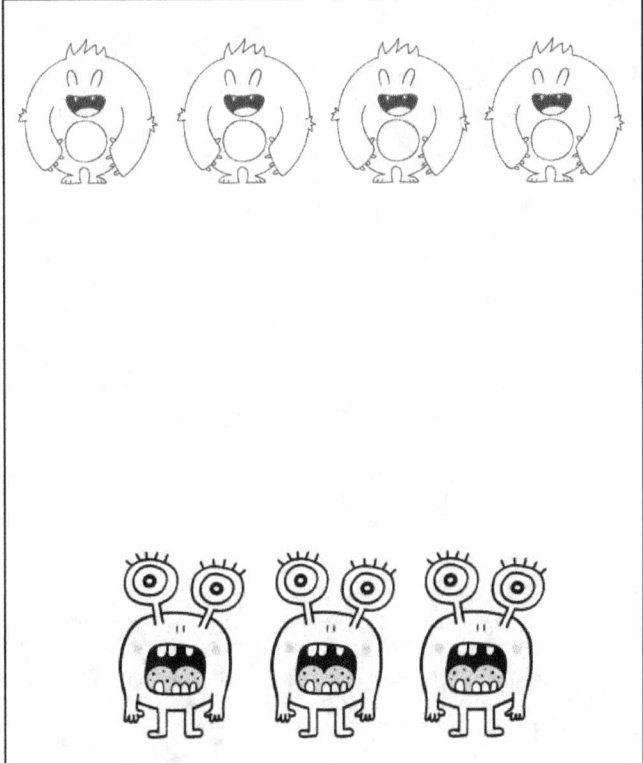
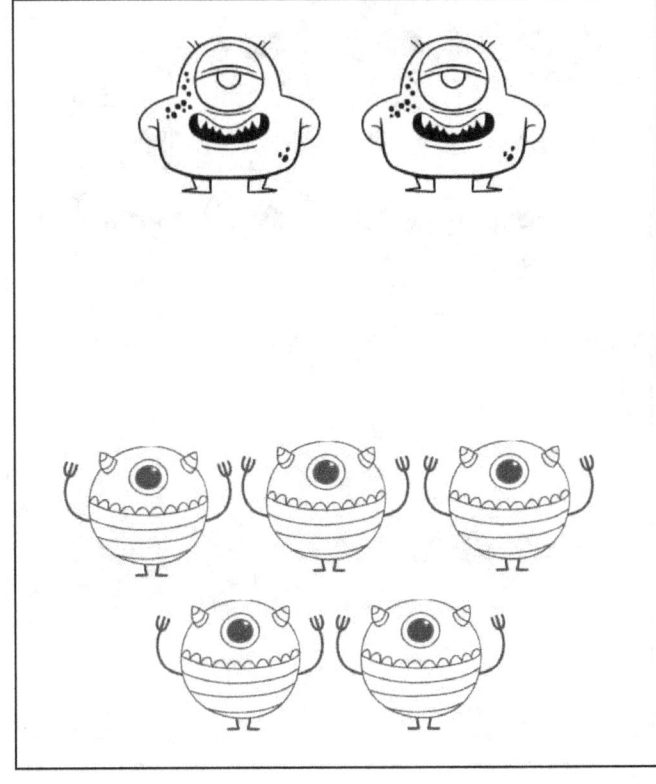

More or Less?

Look at the pictures in each box and circle the group that has **less**.

More or Less?

Count the number of images and write the total number in each box. Circle the box with **more** images in each row.

Total Number:_____   Total Number:_____

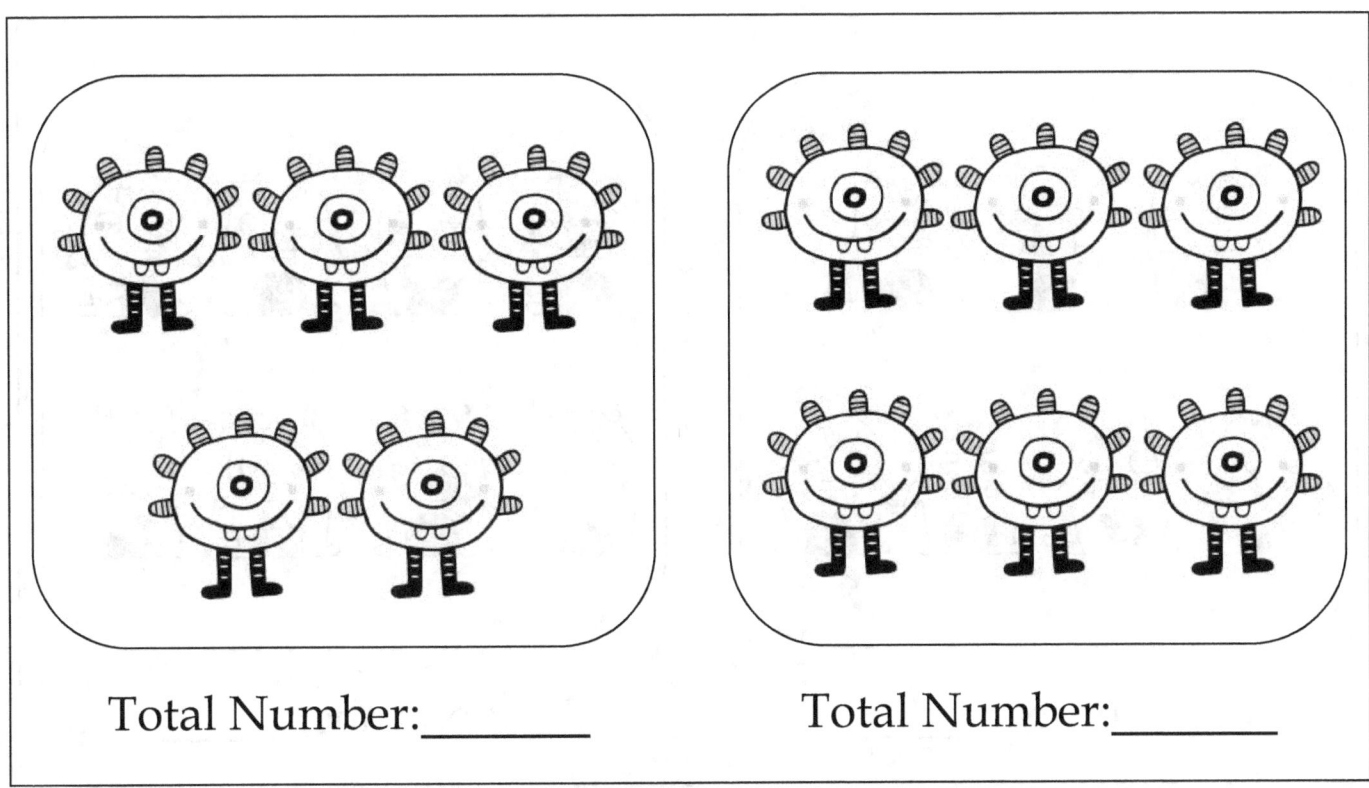

Total Number:_____   Total Number:_____

More or Less?

Count the number of images and write the total number in each box. Circle the box with **less** images in each row.

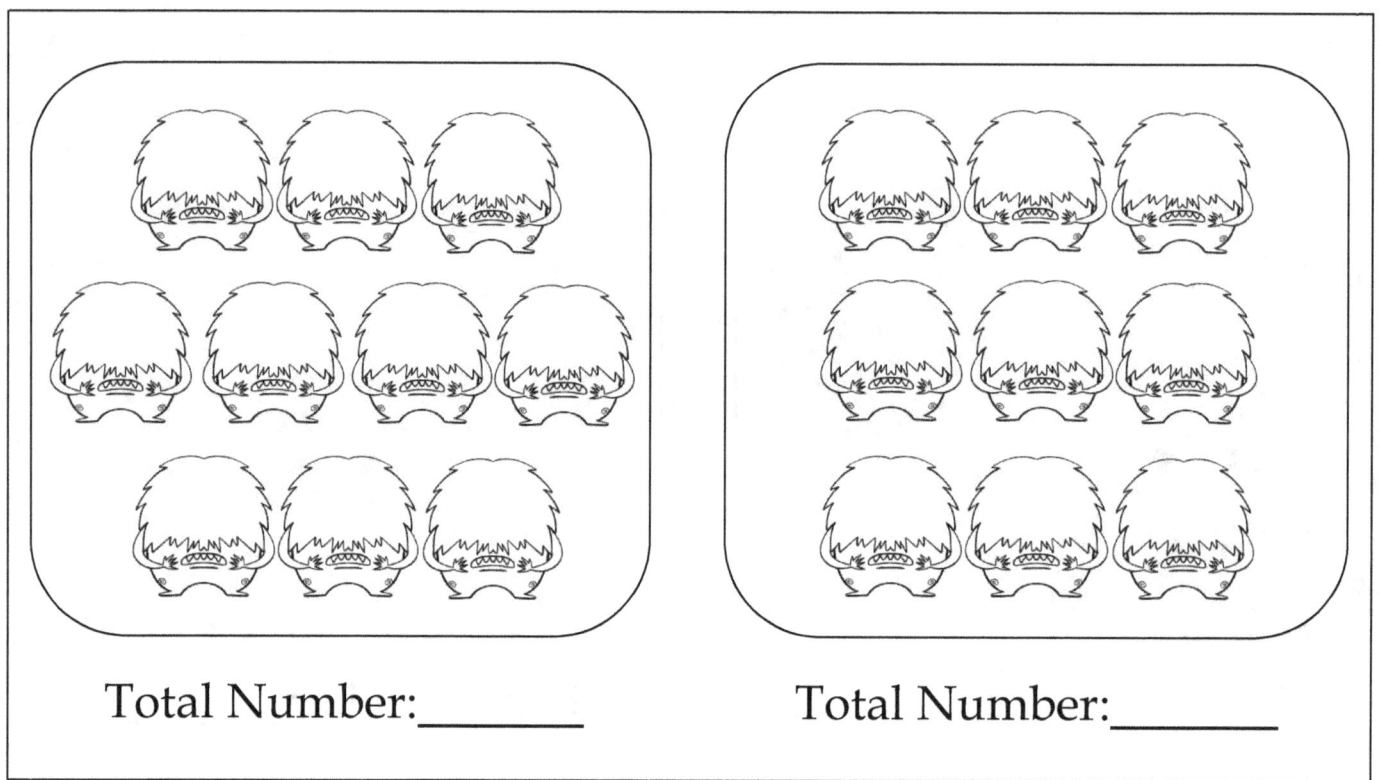

Total Number:_____   Total Number:_____

Total Number:_____   Total Number:_____

Big vs. Small

Look at the picture below and answer the question by circling the correct picture.

Which is **bigger?**	Which is **smaller?**

Which is **smaller?**	Which is **bigger?**

Big vs. Small

Look at the picture below and answer the question by circling the correct picture.

Which is **bigger?**	Which is **smaller?**
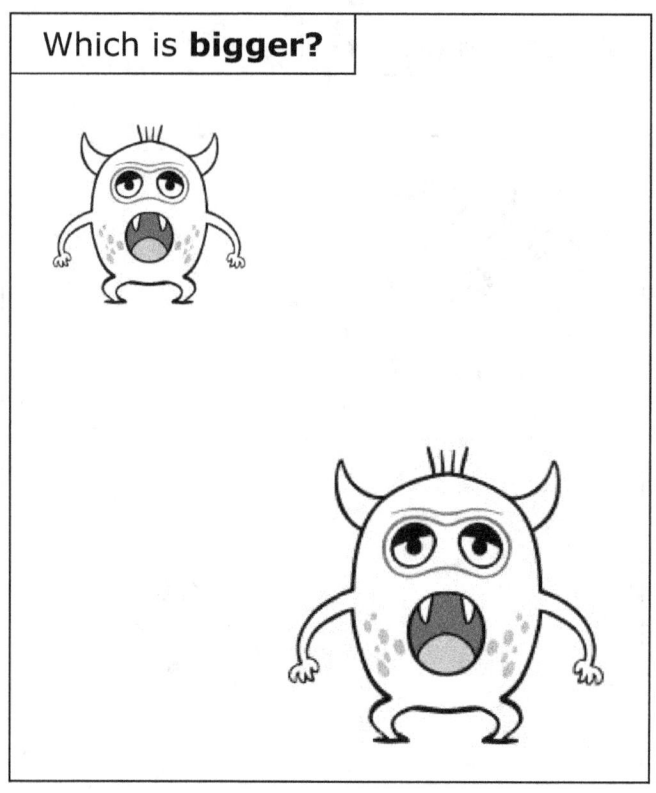	
Which is **smaller?**	Which is **bigger?**

Big vs. Small

Look at the picture below and answer the question by drawing the correct picture.

Draw a **bigger**

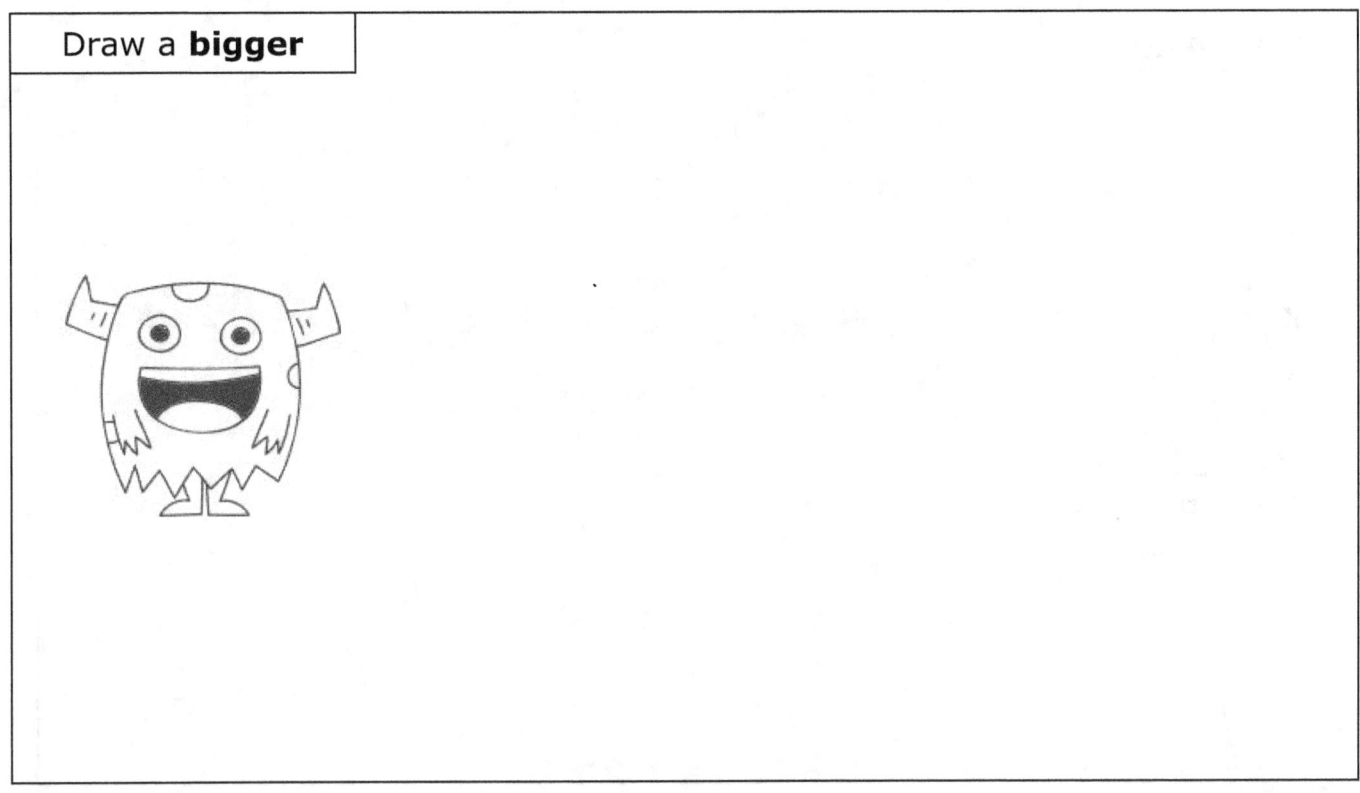

Draw a **smaller**

Big vs. Small

Look at the picture below and answer the question by drawing the correct picture.

Draw a **bigger**

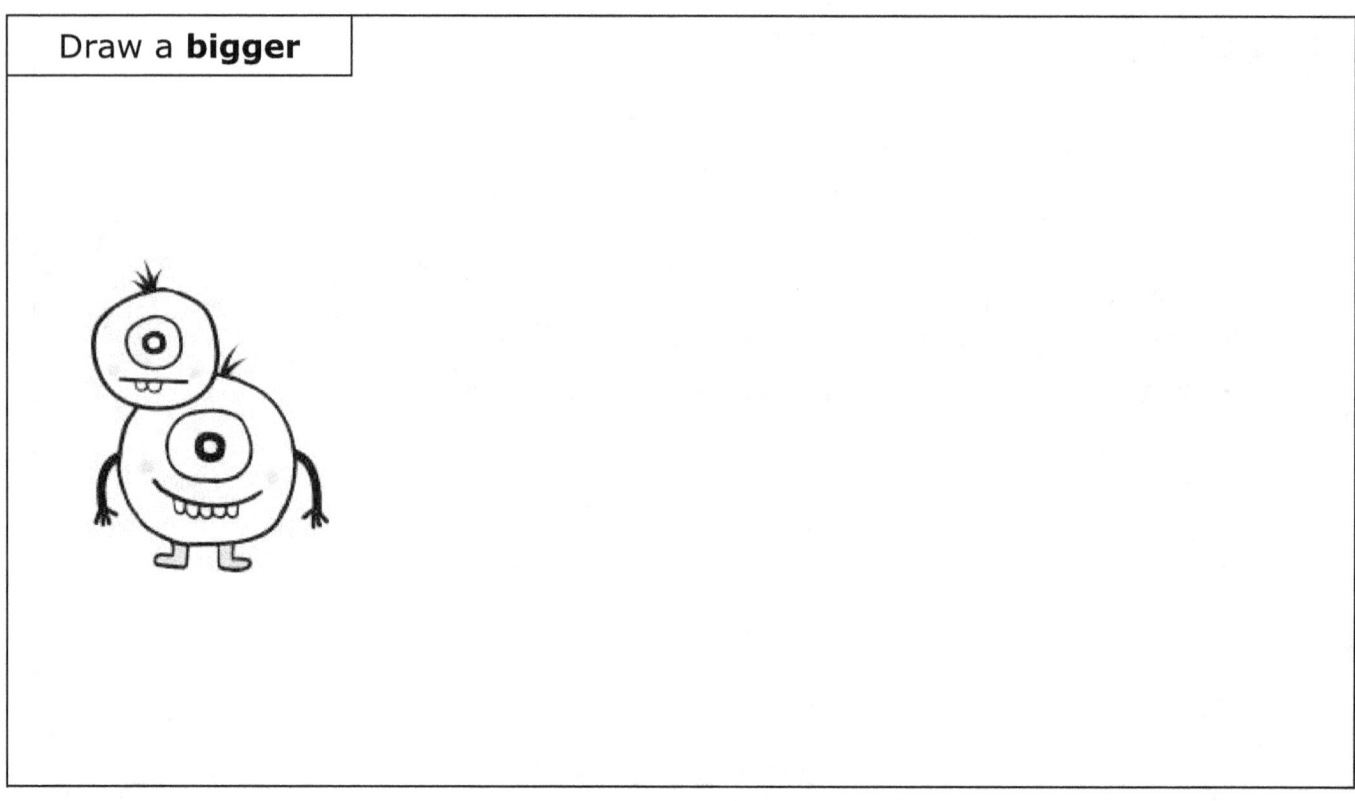

Draw a **smaller**

Big vs. Small

Look at the picture below and answer the question by drawing the correct picture.

Draw a **bigger**

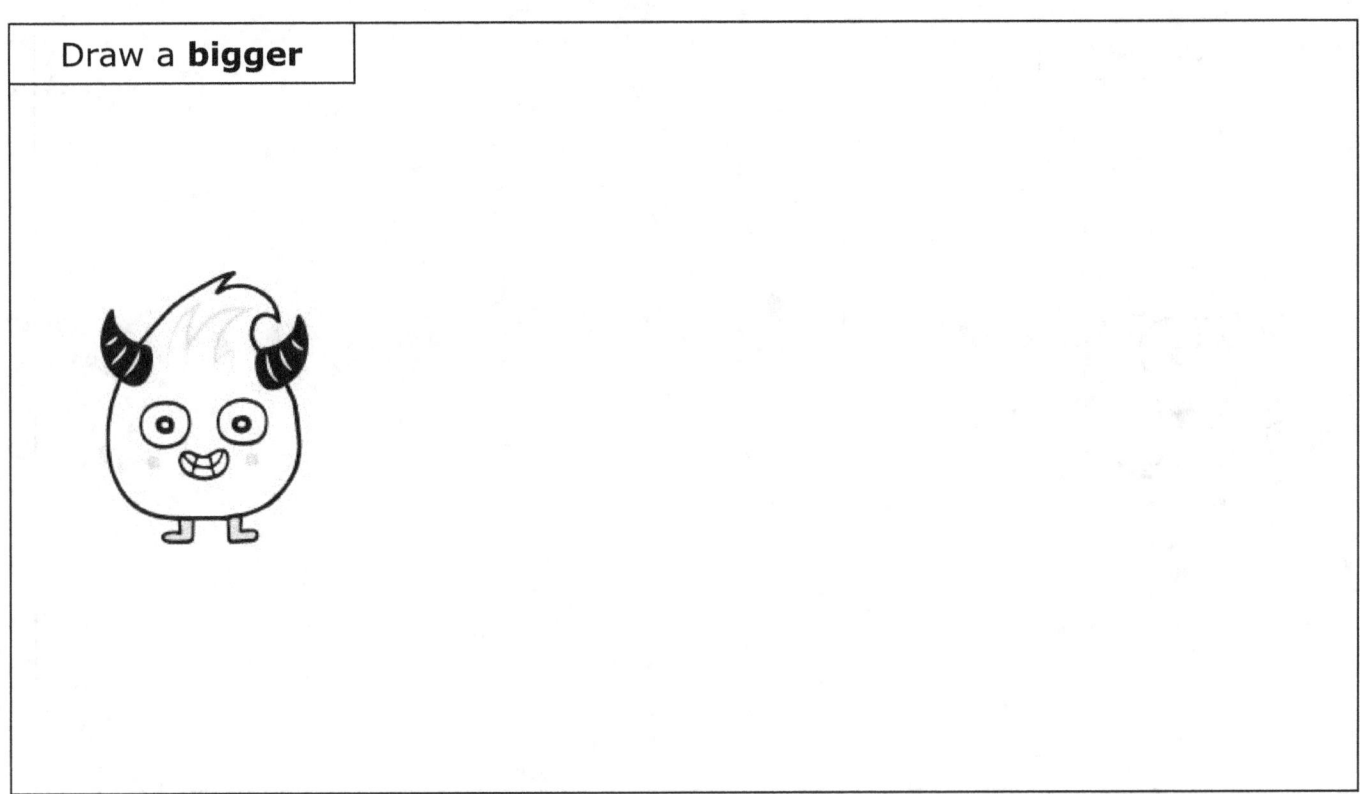

Draw a **smaller**

Big vs. Small

Look at the picture below and answer the question by drawing the correct picture.

Draw a **bigger**

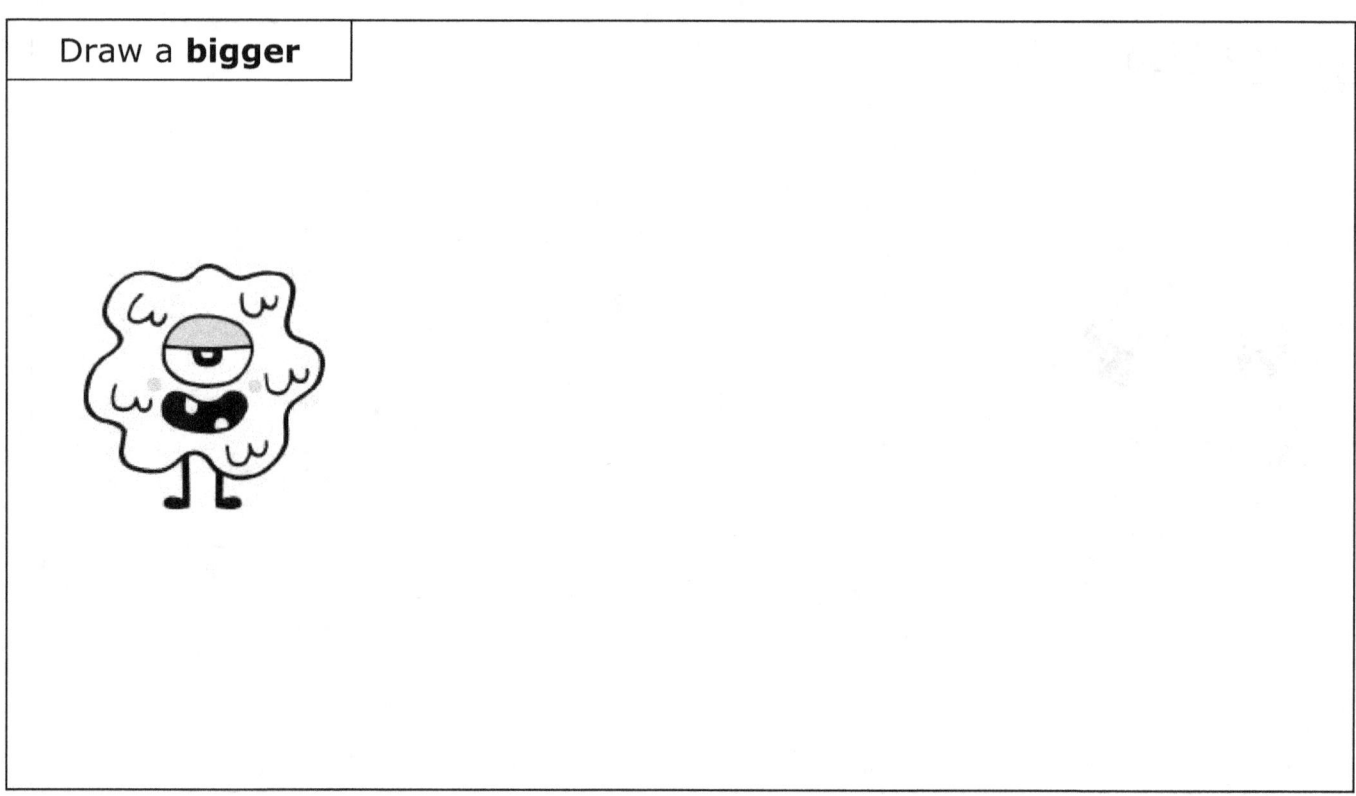

Draw a **smaller**

Compare sizes

Small   Medium   Large

Color the large in red, medium in yellow and small in green

Color the large in red, medium in yellow and small in green

Compare sizes
Color the large in red, medium in yellow and small in green

Color the large in red, medium in yellow and small in green

Shape - Circle

circle

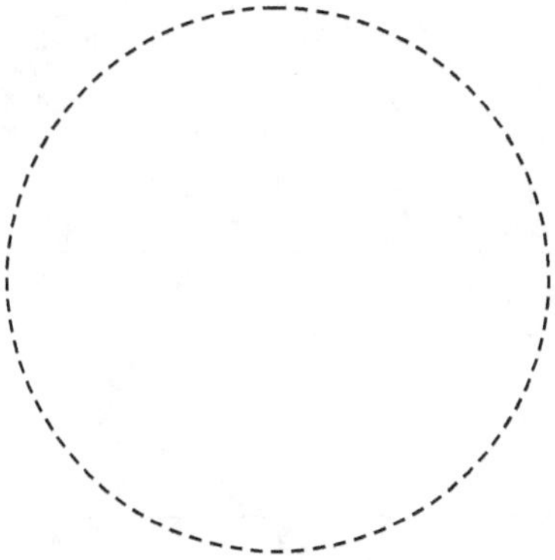

Let's trace the circle!

Let's draw the circle!

A circle has _____ sides and _____ corners.

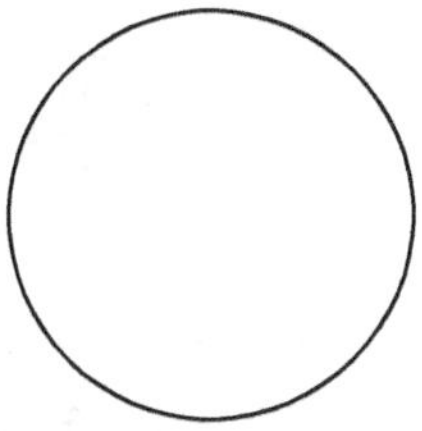

Circles are hiding in the images. Find and color the circles!

Shape – Triangle

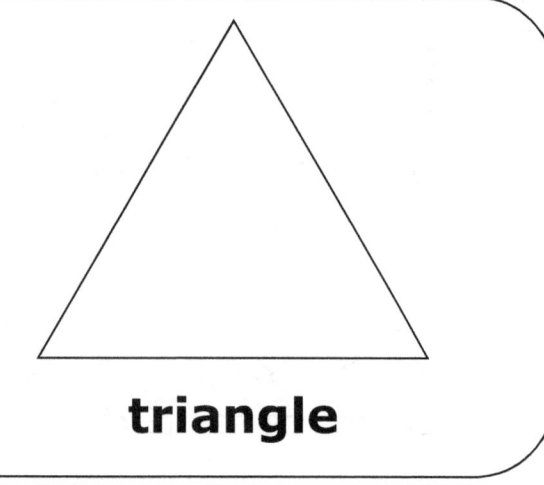

triangle

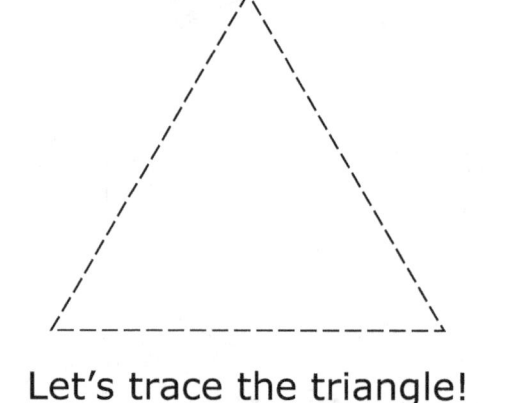

Let's trace the triangle!

Let's draw the triangle!

A triangle has __ sides
And _____ corners.

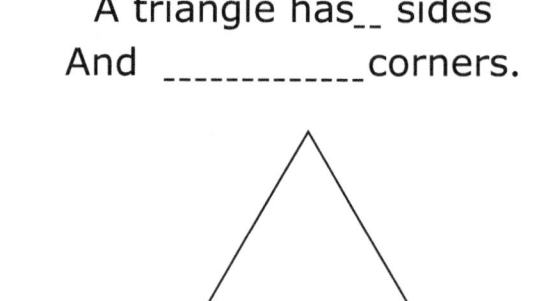

Triangle are hiding in the images. Find and color the

Shape – Rectangle

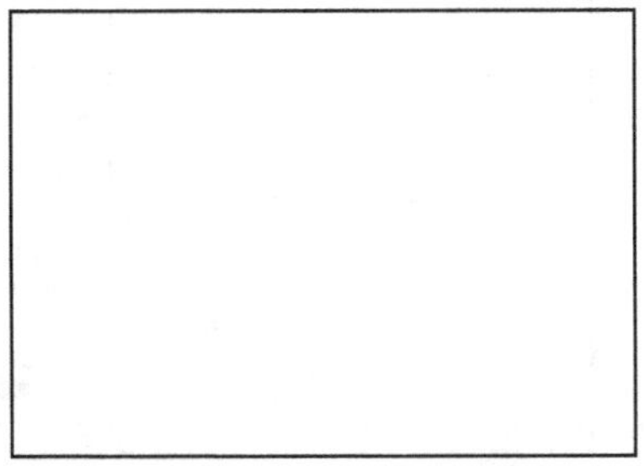

rectangle

Let's trace the rectangle!

Let's draw the rectangle!

A rectangle has_____ sides
And _____ corners.

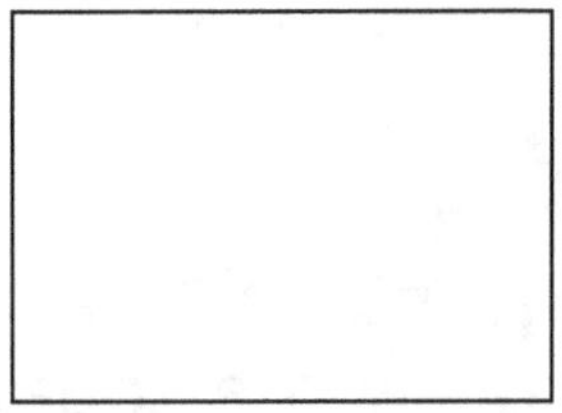

Rectangle are hiding in the images. Find and color the rectangle!

Shape – Square

square

Let's trace the square!

Let's draw the square!

A square has___ sides
And _____ corners.

Square are hiding in the images. Find and color the square!

Shape – Trapezoid

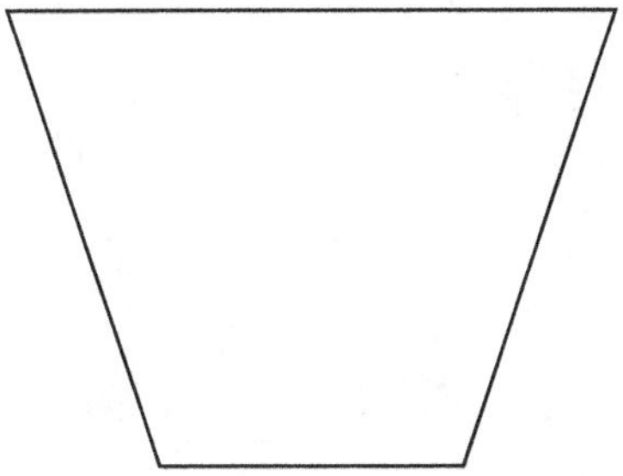

trapezoid

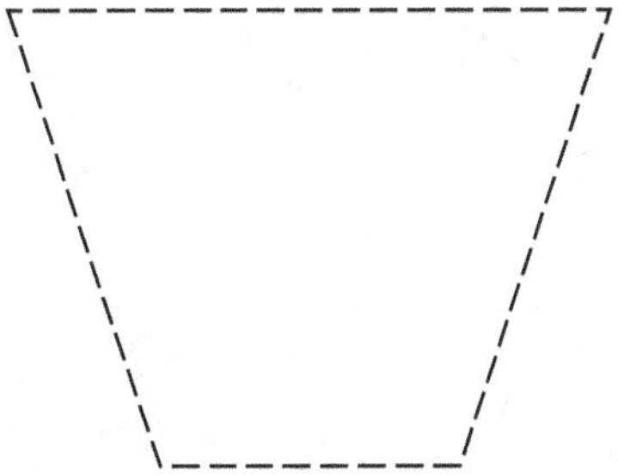

Let's trace the trapezoid!

Let's draw the trapezoid!

A trapezoid has_____ sides
And _____ corners.

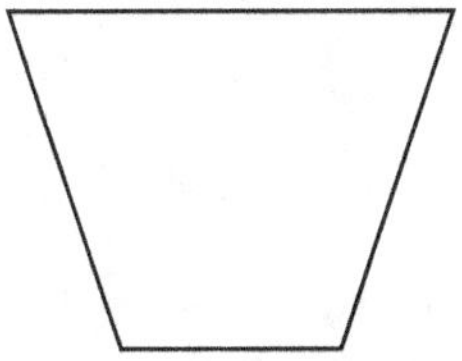

Trapezoid are hiding in the images. Find and color the trapezoid!

Shape – Pentagon

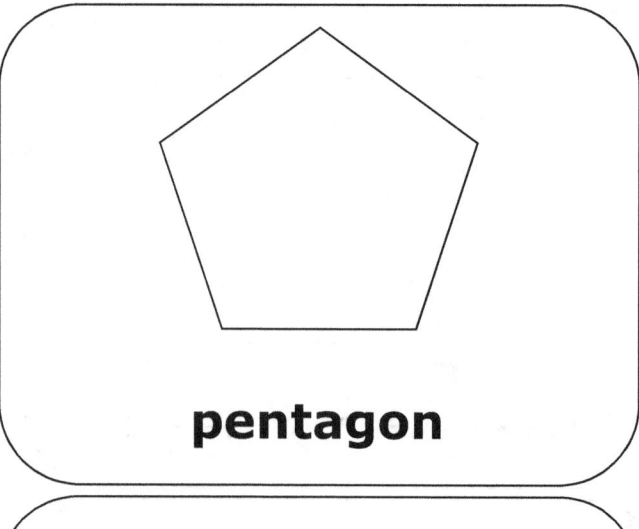

pentagon

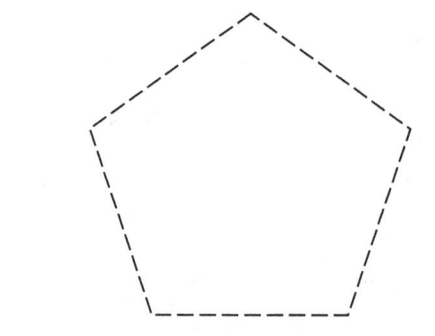

Let's trace the pentagon!

Let's draw the pentagon!

A pentagon has_____ sides
And _____ corners.

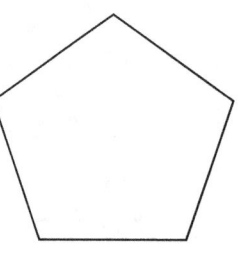

Pentagon are hiding in the images. Find and color the pentagon!

Shape – Hexagon

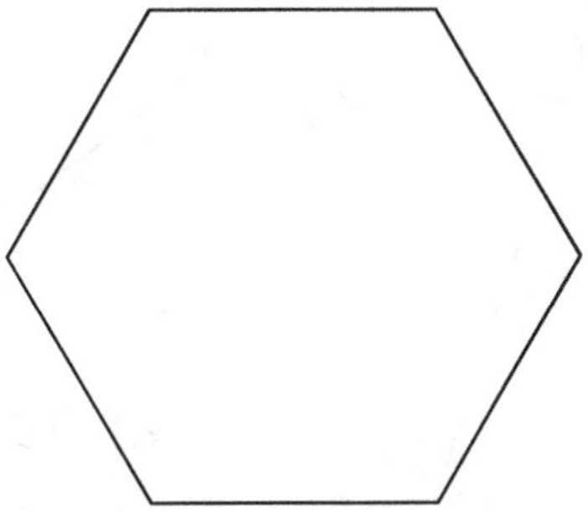

hexagon

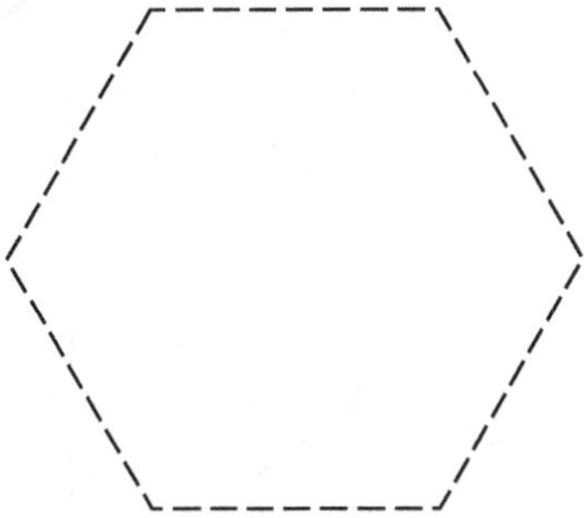

Let's trace the hexagon!

Let's draw the hexagon!

A hexagon has_ sides
And _____ corners.

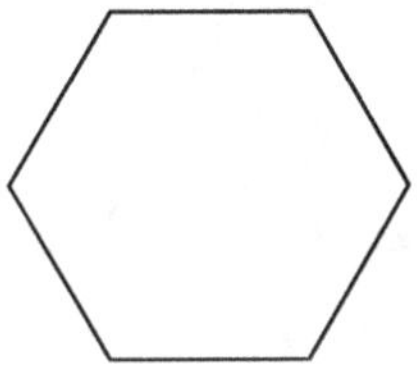

Hexagon are hiding in the images. Find and color the hexagon!

Color the correct number of sides for each shape.

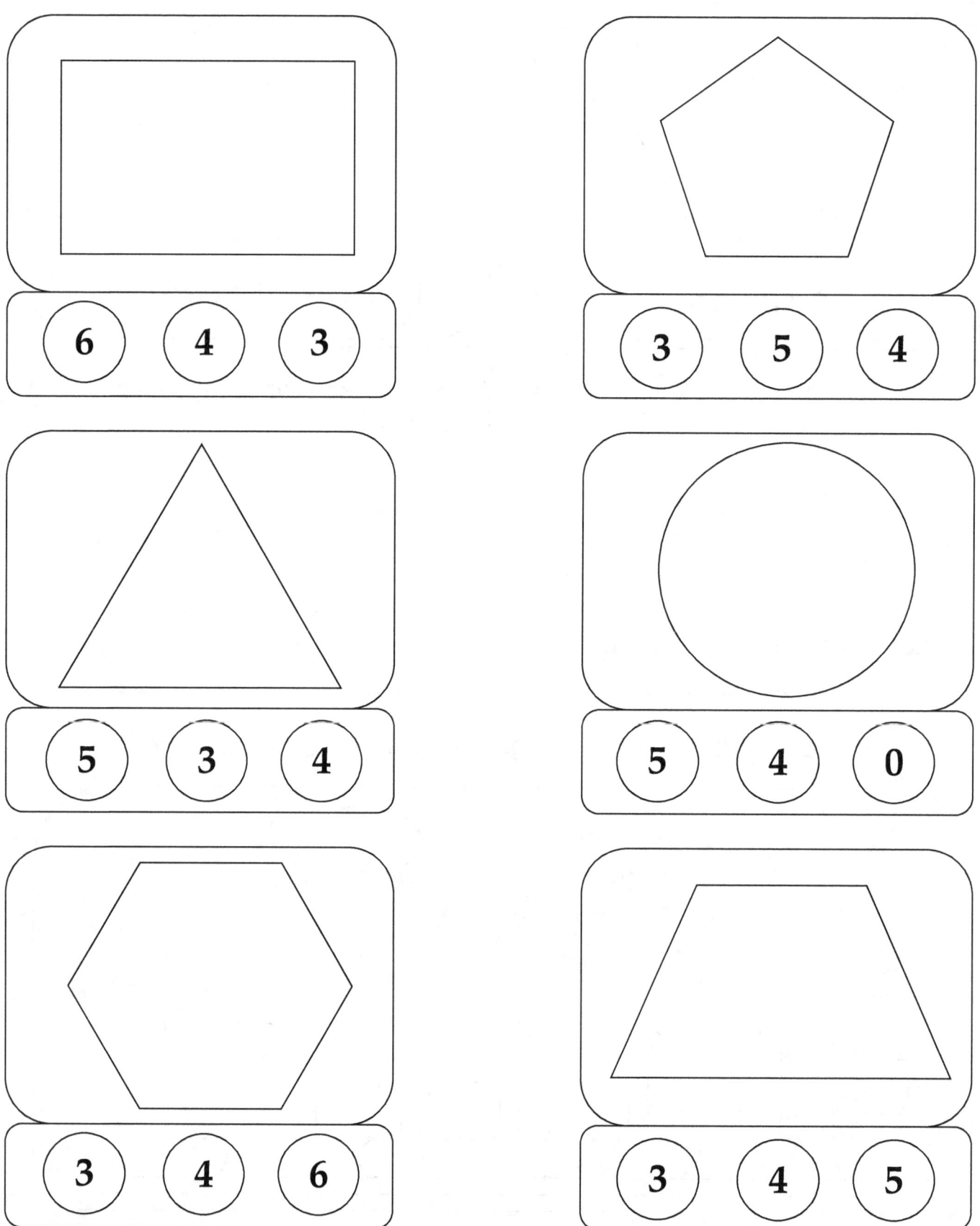

The picture that comes next!

Look at the patterns below. Cut out the images and paste the image that comes last in each box.

The picture that comes next!

Look at the patterns below. Cut out the images and paste the image that comes last in each box.

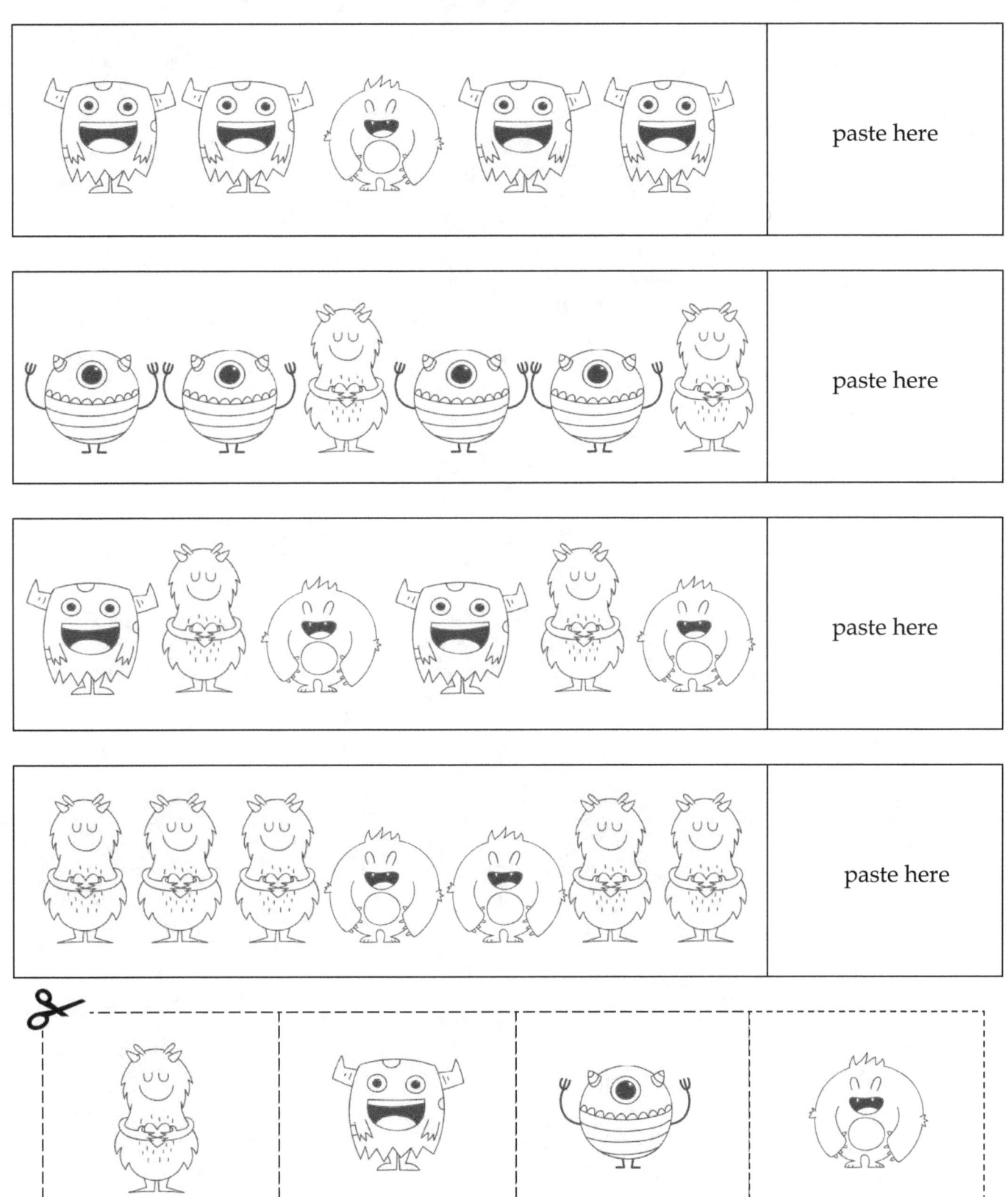

The picture that comes next!

Look at the patterns below. Cut out the images and paste the image that comes last in each box.

Number Fun

Counting up from 1, which number comes next?

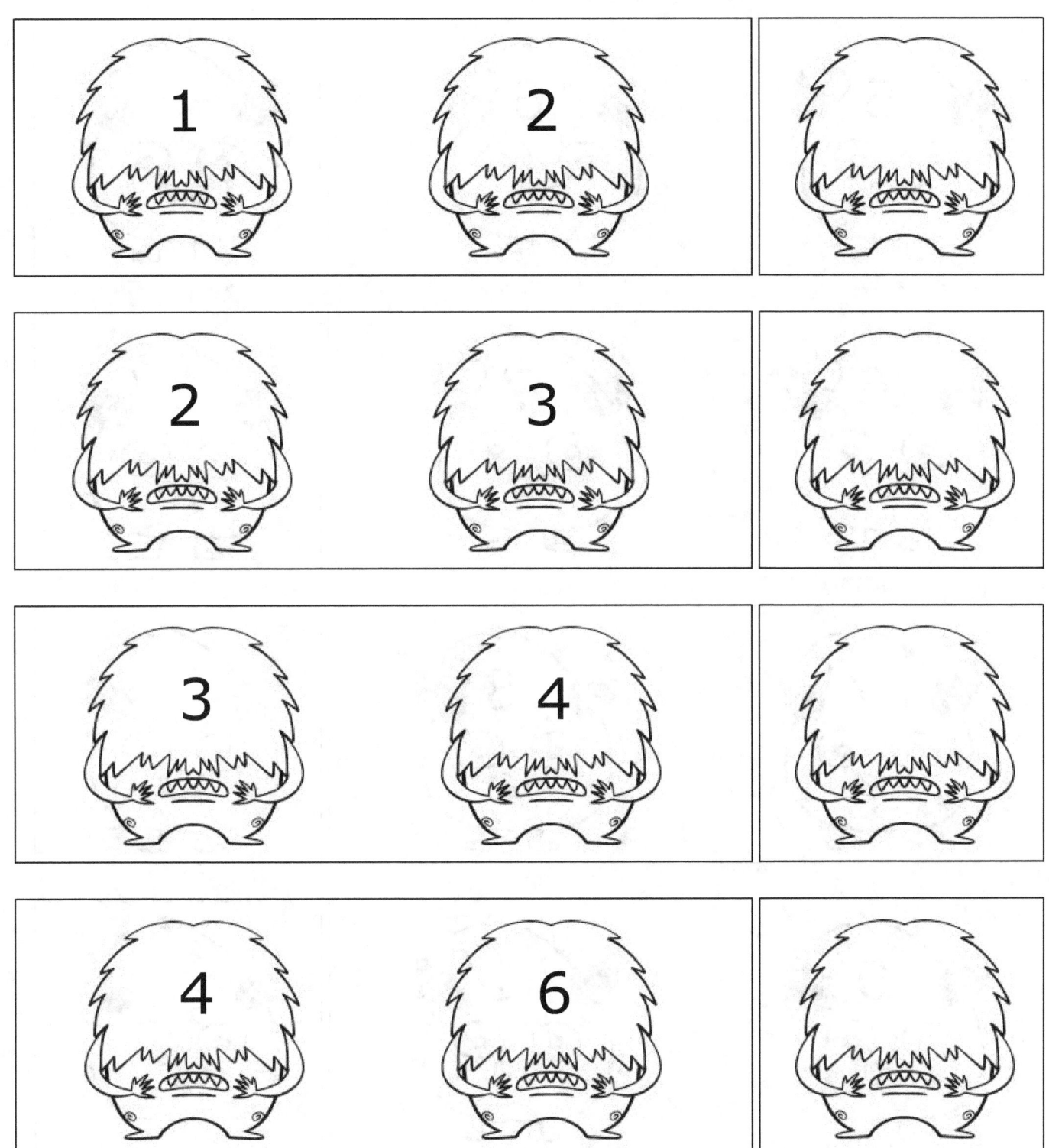

Number Fun

Counting up from 1, which number comes next?

Number Fun

Counting forward from **3**

Counting forward from **4**

Counting forward from **1**

Counting forward from **7**

Number Fun

Counting forward from **2**

Counting forward from **0**

Counting forward from **6**

Counting forward from **5**

Number Fun

Find number in the pictures and compare, then color the one containing the **larger** number.

Number Fun

Find number in the pictures and compare, then color the one containing the **larger** number.

Number Fun

Find number in the pictures and compare, then color the one containing the **larger** number.

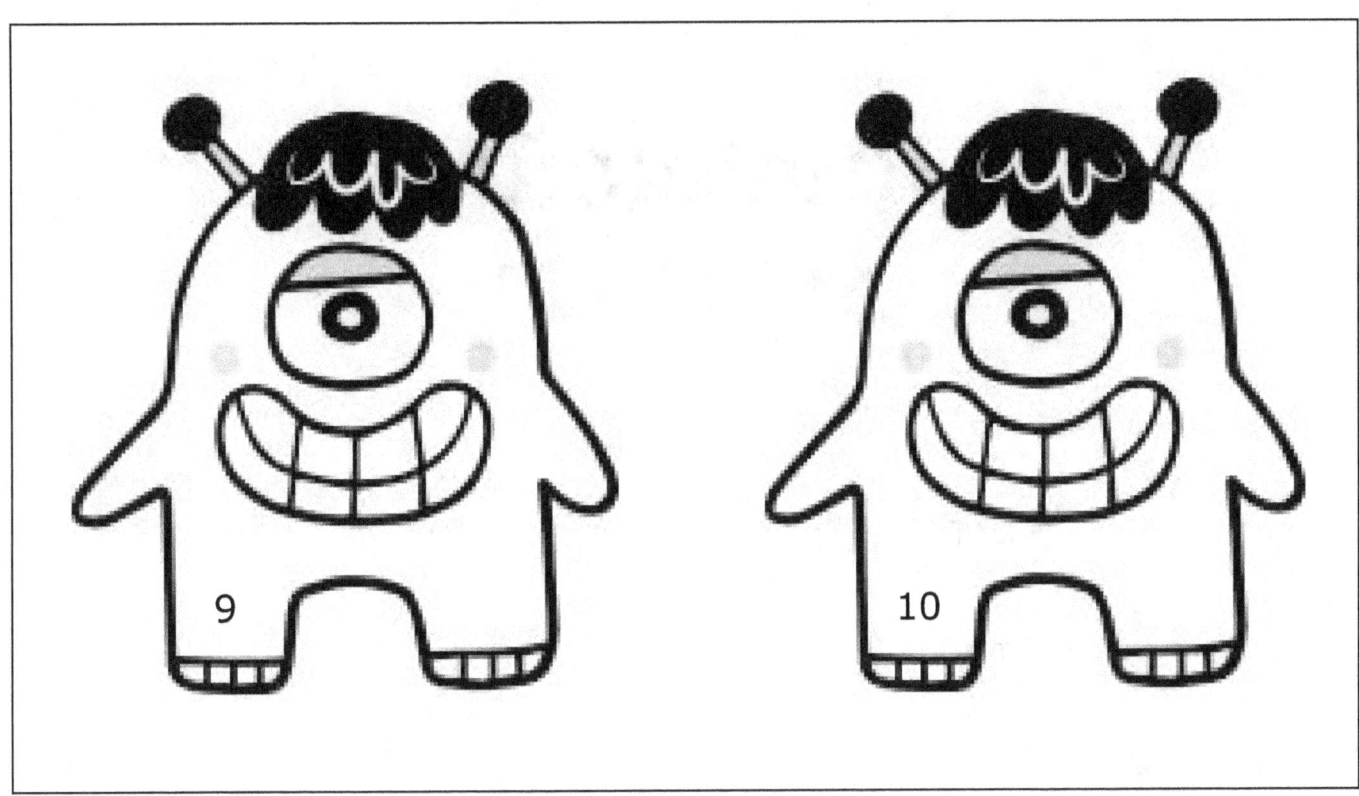

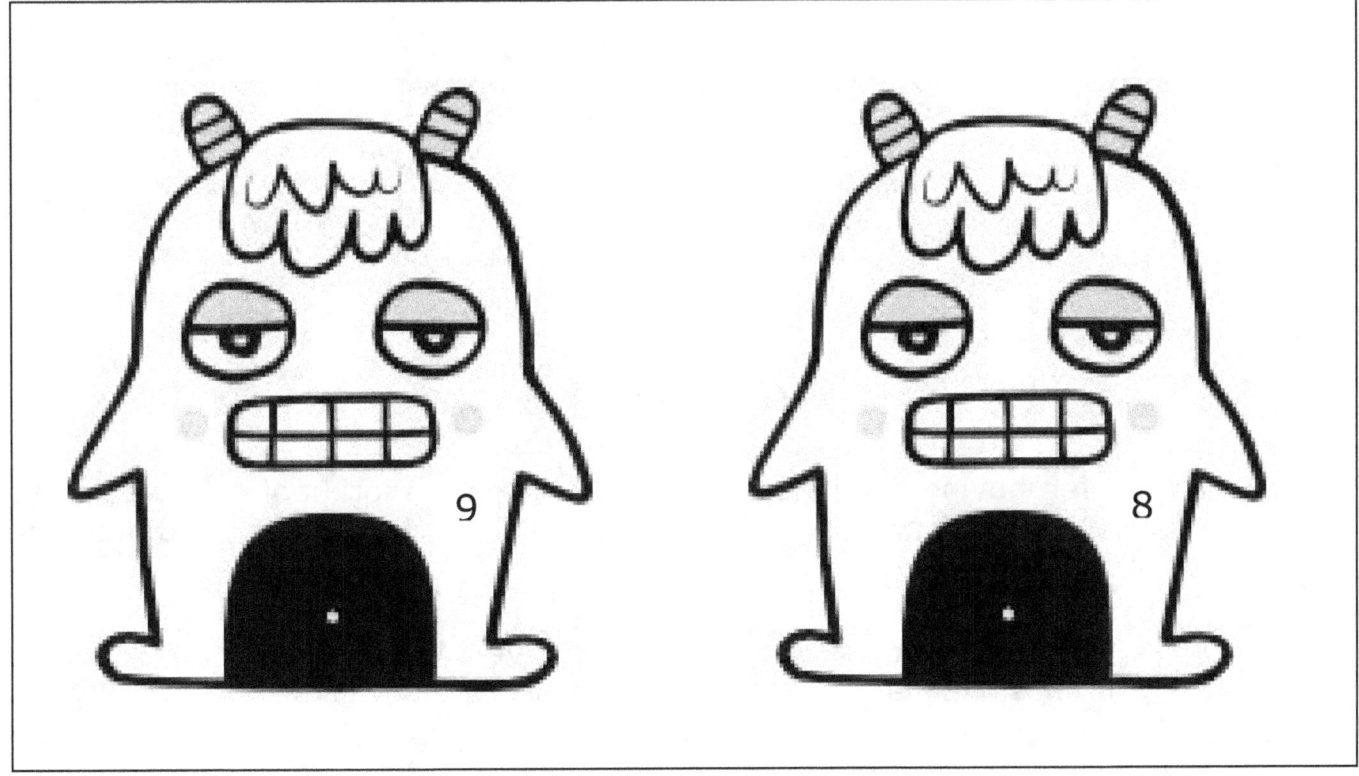

DATE: _____

TIME: _____

Thank you!

Thank you for buying "Monsters" Preschool Basics Workbook we would be more than happy to consider how to apply your suggestion to the next edition. Without you voice, we can't exist.

Please, support us and leave a review!
We welcome your positive feedback and hope that others will benefit from your experience.

www.ingramcontent.com/pod-product-compliance
Lightning Source LLC
Chambersburg PA
CBHW080620220526
45466CB00010B/3404